Jesus and the Future of Transportation

Technology Rooted in Divine Love

Table of Contents

Chapter 1. Introduction

Delving into the unchartered waters of spirituality and technology, our Special Report titled "Jesus and the Future of Transportation: Technology Rooted in Divine Love" uniquely bridges the gap between these two distinct realms. This exploratory expedition will guide you through an enlightening fusion of faith and futuristic travel, demonstrating how the principles of compassion and love embodied by Jesus could play an influential role in shaping transportation technology that is more sustainable, inclusive, and humane. With every turn of the page, feel an enriching blend of theology and technology awaken your curiosity, reminding us all that innovation need not be devoid of love and collective well-being. Ignite your perspective, broaden your horizons, and let us journey together into the heart of this holy intersection, unraveling how divine love could ultimately engineer our transportation future. With anecdotes, expert insights, and thought-provoking projections that are as engaging as they are educational, this report is not only an insightful read - it is a compelling call to action! Don't miss the opportunity to own this unique cross-disciplinary exploration.

Chapter 2. Journey into the Intersection of Faith and Innovation

The evolution of travel is an intriguing subject that converges on the axis of necessity and innovation. As we rev the engines into the future, it's worth halting at the crossroads of faith and invention. This convergence is not merely a haphazard collision of two parallel lines, but an intentional merging of thought processes rooted in the noblest of human faculties - compassion and love, as embodied by Jesus in the Christian faith.

2.1. A Brief Look Back

The harmony of innovation and faith has always shaped our transportation evolution, albeit subtly. Jesus, described as a carpenter, understood and used technology available in his era. The fish and bread, fed to the multitude, isn't just a miracle in religious texts, but an example of how resources can be effectively distributed. The paralleling principles of Jesus's teachings and sustainable development goals extend beyond feeding the multitude to future transportation.

2.2. Sustainability: The Common Ground

As we transcend into smart cities using AI, self-driving cars and globalization, we must pause to consider the impact of our actions on the planet. The concept of stewardship, a significant aspect of Christian teaching, professes responsible management of the world's resources. It ties neatly with optimizing the use of technology in transportation for less pollution, congestion, and economic disparities; creating a vision of a 'green' and 'smart' future, rooted in love and care for our common home.

2.3. Transportation Infrastructure: For All?

Modern transportation's enormity often excludes those deemed 'unfit' to participate by society's norms – be it due to physical disabilities, economic disparities, or age restrictions. The essence of Christian teaching, the parable of the Good Samaritan for instance, sends a powerful message of inclusiveness and kindness towards our fellow travelers on this life's journey. By echoing this in our transit systems, we say that a world moving forward leaves none behind.

2.4. Technology as a Tool of Love

Technology not as the end, but as a means to an end – a tool to further the message of compassion and love. It's by no accident that Jesus chose to be a carpenter, not a warrior. His choice symbolizes the use of tools for creation, not destruction; for love, not hate – paving the way to perceive technology in transportation as a tool to make travel safer, faster, and humane.

2.5. The Ethical Dimension of AI and Autonomous Vehicles

The advent of AI and autonomous vehicles brings forth an unprecedented challenge: Decisions previously left to human judgment are now in the control of algorithms. How can we ensure that these 'intelligent' machines reflect the principles of compassion and mercy? Can AI learn from the parables? Intriguingly, the answer lies in deploying AI not merely as an executor of tasks but as an embodiment of ethical values.

2.6. The Divinity of Innovation – An Unexplored Connection

Miracles in the Bible, like turning water into wine, walking on water, and getting fish and bread for thousands from virtually nothing, are the earliest examples of 'techpreneurship'. These accounts contain invaluable lessons for contemporary innovators. Tech miracles of the Bible set the precedent for use-inspired innovation; they served human needs rather than just showcasing power.

2.7. Technology – Aided by the Spirit of Service

The technology of tomorrow is not just about creating the fastest, the most powerful, or the most advanced vehicles. True innovation embodies the spirit of service. It goes hand in hand with an attitude of humility, not pride; of service, not domination – an attitude that places the welfare of humanity above all.

2.8. Visions for a Future Rooted in Love

Conceptualizing a future where each technological advance in transportation resonates with the principles of compassion and love advocated by Jesus invites us all to play our part in steering this vehicle of human progress. A future that fosters interconnectedness, where journeys are not just about moving from point A to B, but about sharing stories, wisdom, and the joy of being alive.

From wayside Samaritans to self-driving cars, we are continuously reminded that every journey we make is a quest for a deeper connection – a connection that goes beyond logistics and physicality, reaching into the very essence of our being: love. With this thought warming our hearts, let us embark upon the pathway of creation - ensuring that the future of transportation reflects our highest potential, our noblest values, and our supreme aspiration to love all and serve all.

Indeed, let the intersection of faith and technology resemble the intersection of human hearts - always open, always caring, always ready to embrace the other. May every invention echo our interconnectedness, just as Jesus, the carpenter and Messiah, had envisioned for humanity.

Chapter 3. Defining Divine Love in the Context of Technology

Ascertain from the life of Jesus and the teachings of Christianity, divine love is a powerful force that represents the highest form of affection and benevolence. Divine love transcends beyond human comprehension and is often characterized as boundless, unconditional, and self-sacrificing. It seeks the highest good of others, irrespective of one's personal gain or loss.

3.1. Understanding Divine Love

Divine love is often framed within the context of God's unending love for humanity, the manifest essence of which is found in the person and teachings of Jesus Christ. As mentioned in the Bible, "God so loved the world that he gave his only Son, so that everyone who believes in him may not perish but may have eternal life." (John 3:16, NRSV). This excerpt deciphers divine love as a sacrificial force that outweighs every constraint.

While human love can be influenced by personal biases, conditions, and other factors, divine love is sacrosanct and unconditional. It offers acceptance without demanding change, provides solace in times of distress, and invokes peace within tumultuous situations.

Divine love isn't merely an emotion or a feeling that can fluctuate; it is a consistent thread woven into the fabric of creation, ensuring the welfare of every being.

3.2. Divine Love and Technology: An Unheard Alliance

Technology, as we know it today, is a tool wielded by humankind for making various aspects of life more effective and efficient. It is a product of our collective imagination, ingenuity, and intellect. But technology, without a moral compass and a guiding light, can often lead to undesirable and devastating consequences. Such disruptive instances led various technologists, scientists, and moral philosophers to advocate for a sustainable, ethical, and inclusive foundation for technology – a foundation rooted in divine love.

Integrating technology with the principles of divine love may initially seem far-fetched. However, the premise becomes increasingly significant when viewed in the context of the complexities, divisions, and challenges that our global society faces.

3.3. Principles of Divine Love in Building Technology

Taking inspiration from Jesus' teachings, we find several principles that can guide the human operation of technology:

1. Unconditional Love: Technologies should be built with the objective to serve all, irrespective of their social, economic, or racial backgrounds, without any prejudice or bias. The aim here is to make the world a more inclusive and egalitarian place.

2. Selflessness: Technological advances must always seek the betterment of humanity, rather than personal gains. This builds a focus on collective growth that fosters community and unity.

3. Sacrifice: Designers and developers must be willing to sacrifice short term benefits for the welfare of humanity. This includes properly assessing environmental impacts, proceeding with

caution, and renouncing any potential harmful technologies.

4. Compassion: Technologies should be solution-oriented, aiming to relieve suffering and enhance the life quality of users. They should be built with empathy, understanding the needs and desperation of those they serve.

3.4. Embodying Divine Love in Technological Innovation

To drive technology towards a divine love-oriented approach, technologists and designers must consider the following facets:

- Users as Neighbors: As Jesus taught to love your neighbor as yourself (Mark 12:31), this principle can be extended to technology by treating users as neighbors. In software design, for instance, a user-centric, empathetic approach should be adopted that seeks to enhance the user experience rather than exploiting their vulnerabilities.

- Inclusion and Accessibility: For technology to reflect divine love, it should be accessible to all, transcending the barriers of language, physical limitations, economic status, and educational backgrounds.

- Ethical Framework: An inherent ethical framework should guide every step of technological advancement - from conceptualization to implementation and usage.

- Balancing Innovation with Preservation: While technology promotes transformation and growth, it should also ensure preservation. This includes respecting privacy, upholding human values, and reducing harm to the environment.

To navigate the intricacies of technology while maintaining a commitment to divine love, constant reflection, and reassessment are required. It calls for a global dialog among theologians, technologists, policymakers, and the public. Through a concerted effort, we can infuse our technology with divine love, transforming it from a mere utilitarian tool to a holistic instrument of compassion, service, and love—which indeed would be a technology mirroring the very essence of Jesus' teachings.

Chapter 4. From Chariots to Hyperloops: Tracing the Spiritual in Transportation

In the annals of transportation history, the human journey has been characterized by a passionate endeavour to circumvent geographical constraints, making travel faster, easier, and more efficient. Yet, it's not just about technological evolution; it's about the underlying spiritual essence that fuels our shared journey towards progress.

4.1. The Chariot: An Embodiment of Divine Awe

While transportation is firmly rooted in the physical realm, the spiritual dimension pulsates beneath, substantially influencing its course. Ancient civilizations were awestruck by the technology of their time – the chariot. The Egyptians, for example, considered the celestial chariot of Sun God Ra, a symbol of divine power and benevolence, inspiring early transportation dreams.

Chariots were not just transportation devices but spiritual vessels, symbolizing divine will and intervention. They served as vehicles of the gods, sharing a spiritual significance to the people and reflecting the inherent link between the heavens and earth. This connection, crucial in shaping their aspiration towards respectful co-existence, emphasized a deeper essence beyond the mere mobility mechanical transportation offered.

4.2. Bridging the Gap: Spirituality in Shipbuilding

The maritime age, a significant era in the human transportation saga, was not without its spiritual ties, either. The ship, a contemporary mechanical marvel, was an object of reverence and respect for seafaring cultures globally. Early Polynesians relied on spiritual navigation techniques, appealing to ancestral spirits and natural signifiers to sail across Pacific waters.

Compassionate co-existence extended to the way seafarers crafted their vessels. Every element, from selecting the right tree to carve into a canoe to the final farewell as it embarked on its maiden voyage, was regarded as a sacred act, a testament to their spiritual connection with the sea and its governing deity. This deep connection with the natural world helped to develop transportation systems that were more sustainable and harmonious with their environment.

4.3. Railroads and Religion: The Intersection of Progress

As we moved into the Industrial Revolution, rail transport dramatically transformed societies and economies. However, the spiritual undertones of transportation persisted, albeit in a slightly altered form. The advent of trains provoked a religious awakening as some saw this technological advance as God's will to unite people and create a more connected humanity.

Religious groups, with their routines and rituals, played crucial roles in bolstering the moral standing of these iron horses, claiming them to be providential gifts. The transcendence of distance enabled by train travel stirred a sense of divine possibility, a spiritual reassurance that irrespective of physical boundaries, interconnectedness was divine will echoing within the world of iron and steam.

4.4. From Highways to High Skies: An Ascend to Transcendence

The 20th century witnessed an unprecedented leap in transportation technology, with the advent of automobiles and aircraft. However, even these modern marvels cross paths with spiritual beliefs. Around the world, the practice of blessing vehicles, offering prayers for a journey's safety, underscores the human need to intertwine the spiritual with the pragmatic.

Air travel, a technological accomplishment that saw humanity challenging the skies, often evokes spiritual reflection. The human ability to gaze at the world from great heights has been interpreted as a divine gift for some, a testament to an inspirational vision of unity, oneness, and global humanity.

4.5. Hyperloops and Divine Love on the Horizon

As we stand at the brink of another revolution with hyperloops and autonomous vehicles, it becomes pertinent to trace back the spiritual imprints that have guided our transportation journey. This technologically advanced phase should be a continuation of divine love, inclusion, and sustainability history has taught us.

Hyperloops promise near-instantaneous transport, exhilarating us with prospects that seem almost transcendent, blurring physical boundaries and altering the perception of space and time. However, as we strive towards this future, it becomes increasingly essential to maintain our spiritual bearings, to ensure these advancements reflect our higher humanistic values.

The technology of tomorrow must be akin to the celestial chariot of Ra, the compassionate Polynesian ship, or the trains that inspired spiritual unity - an embodiment of love, inclusion, and mutual respect between humanity and nature. It should not be just another conduit towards physical destination but a vessel encompassing the spiritual journey we undertake as conscious beings.

4.6. Shared Mobility, Shared Spirituality

The rise of shared mobility, exemplified by bicycle sharing systems and ride-hailing services, poses another intriguing facet of transportation's future. As we share rides, we move beyond individualistic transportation—towards communal connectivity, subtly resonating with spiritual tenets of unity, brotherhood, and love.

This reinvention of mobility could be an act of divine love in itself if it encourages more sustainable use of resources, and enhances social cohesion. The simple act of sharing a ride possibly echoes a sense of community and togetherness prevalent in ancient chariot and ship cultures. It's not just about reaching destinations but also concurrently sharing and influencing each other's journeys - a vivid embodiment of transportation rooted in divine love.

In conclusion, there is a spiritual dimension to the human story of transportation that transcends mere physical mobility. There's a spiritual continuity, a thread of sacred respect, running through our transportation history that we must continue honouring as we craft the future. Future advancements, such as hyperloops and self-driving cars, must not lose sight of this spiritual ethos. By ensuring new technologies conform with principles of compassion, inclusivity, and respect for nature, we can fashion a more sustainable and humane transportation future where machinery aligns with the spiritual, paving the way for a brighter, divine future.

Chapter 5. The Theology of Technology: Where Does Jesus Fit In?

The fusion of theology and technology is a uniquely intriguing prospect, one which raises many questions that have yet to be fully explored. At the heart of this confluence lies a key figure who transcends boundaries and is universally recognized for his teachings of love and compassion - Jesus.

5.1. Jesus and Technology - A Preliminary Overview

In an age of unprecedented technological advancement, one might initially question the relation between a Biblical figure and the realm of technology. However, when viewed through the lens of philosophy and theology, this connection begins to crystallize. If we think of technology as a human creation aimed at enhancing our lives and solving our problems, then it becomes clear that the teachings of Jesus can indeed influence the ways in which we utilize technology for the greater good.

His message, imbued with values of servitude, compassion, and love for all humanity, can guide the deployment of technology towards more sustainable and inclusive outcomes. In this context, it is crucial to understand Jesus not simply as an historical figure, but as a vessel for principles and values that can influence every aspect of our lives, including our interaction with technology.

5.2. Historical Context & Jesus' Teachings

Despite the fact that Jesus lived in an era that lacked the technological sophistication of our times, his teachings encapsulate timeless truths that can guide us in navigating the high-tech world of the 21st century.

Jesus' life and teachings revolved around principles of love, compassion, and humility. He encouraged his followers to love their neighbors as themselves (Mark 12:31), to show mercy (Matthew 5:7), and to embrace humility (Matthew 18:4). Moreover, he advocated for the care of the poor and marginalized (Matthew 25:31-46), and consistently demonstrated a commitment to inclusivity and justice.

These principles provide a roadmap for the ethical use of technology. When configuring new forms of transportation technology, one can turn to these tenets as a guiding light, ensuring that the systems we create are not just efficient and innovative, but also built with care for our fellow humans and the environment.

5.3. Technological Innovation Through the Lens of Jesus

As we seek to build technology that aligns with Jesus' teachings, we may ponder on the question: how can the ethos of Christ be embodied in the development and deployment of future transportation technologies?

Firstly, inclusivity should be at the forefront of design and implementation. This means creating transportation solutions that are accessible for all members of society, including the elderly, the disabled, and those in economically disadvantaged situations. Leveraging technology to ensure that no one is left behind aligns perfectly with Jesus' teachings of caring for the marginalized and the 'least of these' in society (Matthew 25:40).

Secondly, sustainability must be baked into innovation at every stage. Just as Jesus taught respect for all creation, technological advancement should strive for solutions that minimize harm to the environment and promote ecological balance. Shaping our infrastructure and transportation systems in a way that respects the integrity of God's creation is an essential part of a Jesus-oriented technological framework.

Finally, technology should be deployed with a focus on love and service to others. The transformative technology of the future should not be utilized merely for increased efficiency or profit, but also for the genuine betterment of human lives and communities. This aligns with the core teachings of Jesus, who emphasized the importance of loving one's neighbor and serving others.

5.4. Reimagining the Future With Jesus

In a world increasingly influenced by technology, aligning future developments with Jesus' teachings may seem a daunting task. Yet, as followers of Christ, we are called to carry his message into all aspects of our lives.

As we approach the future of transportation, we must uphold the same values and ideals that Jesus taught: love, compassion, inclusion, and sustainability. As innovation continues to reimagine the limits of human capability, the teachings of Jesus offer a foundational ethical framework, reminding us that our actions and creations should privilege humanity and the natural world we occupy.

By rooting technological development in the values taught by Jesus, we can ensure that the transportation of the future is not only efficient and innovative but also loving, inclusive, and kind to our planet. Jesus and technology may originate from different epochs, but their intersection offers us invaluable guidance. As we progress into a fast-paced technologically advanced future, ensuring that our steps are guided by the principles of divine love could prove crucial in charting a positive and humane course.

5.5. Conclusion : The Jesus Factor in Technology

The juxtaposition of theology and technology may appear unfathomable at first, but a closer examination reveals a natural alignment. Jesus, the embodiment of timeless principles of compassion, inclusivity, and love for all creation, has much to teach us about our interaction with technology, particularly as we explore the realm of innovative transportation solutions.

The heart of Christ's message - helping others, loving our neighbor, caring for the environment - provides a solid ethical foundation that can and should inform our technological endeavors. This understanding could lead us to a future where technology, deeply rooted in divine wisdom, not only transports us from one place to another but also propels human society towards higher ideals of love, compassion, and collective well-being. As we journey ahead, hand in hand with technology, we should remember to look back, drawing wisdom from the age-old teachings of Jesus to navigate the path ahead.

Chapter 6. Life's Ultimate GPS: Navigating Future Transportation with Christian Ethics

In the navigational journey towards the future of transportation, the compass guidance of Christian ethics plays a crucial role in determining our path. We find ourselves in an era where dependency on technology has indeed brought us convenience and improved efficiency but, on the other hand, has given rise to concerns about responsible use, equity, sustainability, and communal harmony. This chapter endeavors to explore the impact of Christian ethics on future transportation systems and highlights the value of walking the path of righteousness as we traverse this rapidly evolving tech landscape.

6.1. Ethical Frameworks and Transportation

Drawing upon the Christian ethic of love, known as "agape", we can find foundational morals that will serve as a beacon of light for building our future transportation systems. "Love your neighbor as you love yourself" is a fundamental commandment in Christianity that underscores the importance of empathy, inclusivity, and a caring society. This idyllic teaching can be purposefully translated to an ethos that technology innovators in transportation must embrace.

Urbanization and increased mobility demand has led to a surge in the development of smart traffic grids, autonomous vehicles, drones, hyperloops, and other modes of high-speed transport that seek to ease congestion and bring planetary advantages. However, multi-fold issues such as accessibility for individuals from various socio-economic brackets, protection of user data, prevention of misuse, ensuring safety, noise and emission limitations need to be addressed through the lens of Christian ethics. By deeply embedding these principles in their architecture, transportation systems can exhibit a genuinely profound agape-inspired love towards society.

6.2. Inclusionary Neighborhoods and 'Agape' Principles

If the core principle of "agape" is intrinsically implemented, technology developers would focus on creating transportation systems that are not only efficient but also perfectly all-inclusive, offering equal accessibility to everyone, irrespective of their income, age, mobility, or geographic location. The principle of justice, along with compassion, could be adopted to ensure that the disadvantaged are not marginalized but have access to affordable, reliable, and decent forms of transport.

In fact, successful models of this inclusion can be found in some European countries, where the focus isn't merely on developing advanced transport, but also on the formation of inclusive neighborhoods that offer essential services within cycling or walking distances.

6.3. Sustainability: The Biblical Principle of Stewardship

In addition to promoting well-being, Christian ethics also highlights the importance of handling God's creation with humility and respect. What does this translate to in terms of transport technologies? Simply put, sustainable modes of mobility that respect the environment instead of exploiting it. Proponents of green transport could find an ally in Christian teachings, with ethical considerations potentially spurring substantial changes in how we approach travel and mobility.

Automated electric vehicles, fuel cell technology, regional rail systems, and other non-polluting modes of transport with low carbon footprints would be fostered under this ethic. Leveraging renewable energy sources to power both personal and public modes of commuting could be seen not just as a technological challenge, but also as a moral imperative.

6.4. Ethical AI Principles for Autonomous Vehicles

Christian ethics speaks volumes about fairness, integrity, and respect for life. As we move towards a future where autonomous vehicles could become the norm, the ethics of artificial intelligence must intertwine with Christian teachings. It is inevitable that autonomous vehicles may face scenarios where they must select between the lives of the passengers, pedestrians, or other drivers.

Guiding principles from Christian ethics, especially the sanctity of every single life, can be embedded into the AI design process to develop decision-making algorithms that prioritize safety and strive to prevent harm at all costs. Transparent and responsible AI, conforming to Christian virtues, would increase public acceptance and participation.

As we continue to embark on this transformative journey, let's consider technology as a tool to implement the divine commandment of love and respect towards ourselves, our neighbors, and our environment. This philosophical shift – or, perhaps, a return to ancient wisdom – can help us make sense of a rapidly digitalizing world, guiding us to a future in which transport is more than just efficiency and speed – it's about shared stewardship and an embodiment of divine love. This is our potential path as we venture onwards, forever aided by life's true GPS - our Christian ethics.

Chapter 7. Paving the Road to Future Transportation with Stones of Compassionate Design

In the grand scheme of technological advances, the paradigm of transportation remains a key component. The rapid evolution of electric vehicles, self-driven cars, hyperloop technologies, and electric air travel denotes mankind's ceaseless aspirations. Perhaps, it's time to gaze upon these innovations from the lens of divine compassion, drawing intrinsic principles from the teachings of Jesus. By figuring out 'the stones of compassionate design', we may lay down a path where the essence of humanity blends beautifully with burgeoning technology.

7.1. The Philosophy Behind Compassionate Design

The teachings of Jesus highlight compassion, love, and service towards fellow beings. This philosophical approach might seem detached from technological development; however, intertwining them could yield surprisingly positive outcomes. Compassionate design takes into account humanity's collective need and employs empathy as its core tool, expecting to curtail any possible technologic repercussions while enhancing its positive impact.

Tech visionaries are encouraged to implement compassionate design that involves considering environmental impact, psychological effects, accessibility, affordability, and cultural harmony. The approach stands at the intersection of technologic pragmatism and empathetic design, intending to generate impactful and sustainable solutions.

7.2. The Fundamentals of Compassionate Design in Future Transportation

Compassionate design underlines five main principles when applied to future transportation: sustainability, inclusivity, accessibility, affordability, and safety. Each principle serves as a 'stone' in paving the path for advanced transport systems.

Sustainability:

Renewed interest in electric vehicles and burgeoning research on alternative fuels reflects the urgency to reduce carbon footprints associated with traditional transportation means. However, the compassionate design tailors beyond just 'green' technology. It strives to optimize resource use, reducing waste and promoting recycling, placing sustainability at its core.

Inclusivity:

A transport system rooted in compassion ensures everyone's access irrespective of age, class, race, or disability. The usage of universal design promotes inclusivity by creating comfortable, easy-to-use systems for all.

Accessibility:

Similar to inclusivity, a compassionate transport system ensures easy accessibility for everyone. It focuses on eliminating transport deserts, improving connectivity within remote or underprivileged areas, thus annihilating any geographic disparity.

Affordability:

The compassionate design emphasizes developing transport modes that are economical, thus ensuring everyone's utilization without the barrier of high costs.

Safety:

Transport safety is paramount. The emphasis lies in creating systems that ensure the safety of its users, accommodating all possible concerns from traffic accidents to cybersecurity for autonomous vehicles.

7.3. Building Blocks of Compassionate Transportation Design

Moving beyond the principles, the actual 'stones' of compassionate design are the individual policies, research, prototypes, and initiatives reflecting these principles.

Policies encouraging research and development for alternative fuels, recyclable parts, and an optimized supply chain should be a welcome move. Economic incentives aimed at 'greening' the current transport infrastructure can also offer strong impetus.

Initiatives focused on inclusivity can entail designing public transport with automated ramps for physically disabled individuals, or GPS-enabled, voice-controlled buses that assist the visually impaired, bridging the inclusivity gap.

The future of transportation must strive to cover all areas, including urban, rural, and remote regions, thus offering universal accessibility. Innovative solutions could be drones delivering essential goods to remote regions, or an enhanced rural public transportation network.

Lastly, investment into safety technologies is absolutely crucial, making travel safe to the level of being foolproof. Autonomous vehicles should ensure top-notch cybersecurity systems, and public transport should imbibe advanced safety mechanisms.

7.4. Bringing Jesus' Teachings to Future Transportation Design

While efforts to improve the empathetic aspect of transportation technologies are underway, it's essential to derive the essence of compassion directly from the teachings of Jesus. These teachings can be boiled down to 'do unto others as you would have them do unto you', also known as the Golden Rule.

This golden rule acts as the guiding principle when developing new transportation advances. If we desire a clean environment, affordable services, easy accessibility, inclusive facilities, and safe rides, shouldn't we design the same for others?

The crucibles of Jesus's teachings, from the Sermon on the Mount to His parables, resound the values of love, generosity, and fairness. We can no longer afford to segregate spirituality and technology. We need to amalgamate them compellingly so we design paths of transport that serve the collective good of humanity.

7.5. Conclusion: Comprehensive Compassion is the Way Forward

While we dream and design the future of transportation technology, let's not forget to merge it with compassion's power. By doing so, we embrace an inclusive, accessible, and affordable future for transport, moving past conventional norms and displaying how divine love can help engineer a transportation future that's reflective of our collective humanity.

This fusion of spirituality and technology is only the beginning of a journey that requires persistent exploration. It's the call to place humanity at the heart of innovation, where each development, each discovery, and each invention aligns with the age-old virtues of love, compassion, and service.

This idea may seem ambitious, perhaps even idealistic. But isn't that what progress and innovation are all about? As our knowledge and skills evolve, so should our ideals and aspirations. For the ambition of yesterday is the reality of today, and maybe the future of compassionate design in transportation. Our technological advancements should not and must not outpace our spiritual growth. In merging the two, we will find true progress. In the compassionate design of our transportation future we are embracing not only the teaching of Christ but also cherishing the deep-seated values of humanity. Let us pave this divine road ahead, together.

Chapter 8. Sustainable and Inclusive: Future Transportation in the Light of Divine Love

The philosophy of Jesus was rooted in profound compassion and love for all, emphasizing the importance of equality and brotherhood. As we look toward the future of transportation, it's hard to recount the manifold ways that these principles could play a crucial role in our advancement. Building a sustainable and inclusive transportation system is a challenge that requires not just technological innovation, but also a reimagining of our values and attitudes at its core. Let's delve into these ideas, keeping emphasis on how we can inherit the divine principles of Jesus to shape our transportation future.

8.1. Mending the Divide: The Concept of Inclusive Transportation

The crux of inclusive transportation is in creating mobility solutions that are accessible to all, regardless of their physical ability, socioeconomic status, or geographical location. Today, a significant faction of our society feels left out, marginalized, or ill-served by the existing transportation infrastructure. According to the World Bank, only 40% of the population in low-income countries has access to affordable and efficient public transportation.

Jesus preached inclusivity, extending his message of love and care without judgment or discrimination. In the Gospel of Luke, Jesus mentions, "He causes his sun to rise on the evil and the good and sends rain on the righteous and the unrighteous." This profound message transmits the essence of inclusion - everyone deserves the right to access resources regardless of their circumstances. By rooting our transportation policies in this principle of universal love, we can begin to address the disparities that characterize our current systems.

8.2. Sustainable Progress: Harnessing Divine Love to Protect the Planet

Environmental sustainability has become a pressing issue in the current age, and transportation is one of the largest contributors to climate change. A study published in the Journal of Industrial Ecology reveals that transportation accounts for nearly 30% of all global energy use and 23% of the world's Green House Gas emissions.

In the Bible, Jesus showcased a deep reverence for the environment. The book of Genesis posits the Earth as a divine creation that needs our care and respect. To align our transportation systems with this ethos of environmental stewardship, we must divert our focus towards sustainable solutions. We must prioritize energy-efficient methods, advance in renewable fuels, and contribute towards a systematic decrease in emissions.

8.3. Multi-Modal Integration: The Path Forward

Often, people living in rural areas find it challenging to access urban services due to the lack of efficient transit systems. Jesus' principle of service towards all necessitates that we build a transportation ecosystem where every region and individual is catered to, irrespective of the density of population or geography.

Multi-modal integration can be a solution. It refers to the provision of a seamless and connected transportation experience by integrating different modes of transport. With a focus on connectivity, inclusivity, and convenience, multi-modal integration accelerates our journey towards a transportation infrastructure that embodies Jesus' teachings.

8.4. The Role of Technology in Forging an Inclusive, Sustainable Future

Emerging technologies offer unprecedented opportunities to overcome many of the current challenges in transportation. Self-driving vehicles can provide mobility for people who are unable to drive. Drone deliveries could bypass traditional infrastructure limitations, reaching remote areas effectively. Artificial Intelligence could optimize public transport routes, making them more efficient and accessible.

It's essential, however, to ensure these technologies are guided by the principles of love and inclusivity exemplified by Jesus. Development should be aimed at serving humanity as a whole, and not just a privileged few. It is equally vital to verify that technological progress does not come at the expense of the environment or the marginalization of any section of society.

8.5. Conclusion

In conclusion, the pathway to a sustainable and inclusive future of transportation, when guided by the divine love that Jesus championed, becomes a journey of collective compassion, inclusivity, and love - a journey that values not just progress, but the very essence of humanity itself. This vision extends beyond mere words; it provides the bedrock for a transformation that can shape our societies for the better.

As we continue to innovate and evolve, let us remember that at the heart of every technological leap, there must be a core of divine love and goodwill – because in the end, our technologies are only as good as the values we instill in them.

Chapter 9. Dystopia or Utopia? Christianity's Imprint on Future Transportation

At the dawn of this new technological era, humanity stands on the cusp, looking forward into a future that is as exciting as it is uncertain. Will we march inexorably towards a world shaped by cold efficiency and impersonal automation? Or can we carve a path that pioneers new frontiers of transportation while preserving the warmth of human empathy and compassion?

9.1. The Shape of Things to Come

The shape of forthcoming transportation systems highly depends on the factors driving their design and implementation. The current trend suggests a significant shift towards automation, electrification, and digitalization, spearheaded by an eager desire for efficiency, profitability, and environmental sustainability.

Why discuss these primal forces shaping transport's future through the lens of spirituality, specifically Christianity? One might argue it offers a drastically different perspective, reminding us that technology need not only be a tool for efficiency but also an extension of our commitment to care, integrity, and love for one another.

9.2. The Dystopian Vision of Future Transportation

If we let these forces alone shape our future, a potentially dystopian vision emerges. High-speed hyperloops and supersonic planes may promise unparalleled efficiency, but, devoid of a human touch, they risk isolating individuals, reducing shared journeys to mere transactions and turning transportation into an impersonal, automated process secondary to the thrill of arriving at a destination.

Moreover, the 'survival of the fittest' mentality underpinning the current competitive landscape may create a divide between haves - those who can afford the latest technologies and autonomous vehicles, and have-nots - the ones left behind in the race, unable to access these developed technologies due to financial, physical, or cognitive constraints.

9.3. The Utopian Vision: A Message of Love

However, by contrast, Christianity offers a different viewpoint – a utopian vision for future transportation. At its core, Christian teachings preach compassion, humility, and love for one another. If we integrate these principles into the framework of future transport - striving to build systems that are not only efficient but also inclusive, considerate, and compassionate - we could create a world where technology serves as an extension of divine love.

In this utopia, transportation is not merely about moving bodies from point A to B, but nurturing connection, fostering interaction, and encouraging empathy. Autonomous vehicles, for instance, can be designed to accommodate the elderly, the disabled, and those with cognitive restrictions, focusing on providing a dignified journey for all, irrespective of their physical or cognitive abilities.

Public transportation could move beyond just efficiency, embracing principles of design that encourage human interaction, shared experiences, and stronger communities. Technological advancements in transport need to honor the environment to ensure our collective well-being, focusing on sustainability as much as speed.

9.4. The Role of Christian Ethics in Future Technology

The Christian principle of sacrificial love mandates caring not only for our immediate interests but also for those of others. When applied to transport technology, this could mean developing systems which prioritize the needs of the community, particularly the most vulnerable, over profit or pure efficiency.

Our systems could honor diversity in physical and cognitive abilities, exemplifying inclusivity. Transportation channels could be designed to become shared spaces of camaraderie and interaction, honoring our need for meaningful connections and shared experiences.

Moreover, Christian ethics mandate stewardship of the Earth, providing a spiritual foundation for sustainability. By embracing green technologies and reducing carbon emissions, we could ensure that our advancements do not destroy our planet but rather, they honor and preserve it.

9.5. What It Will Take to Walk on This Path?

Walking on this path will take conscious effort and collaboration across various sectors. Technologists, policymakers, civil society, and spiritual leaders will need to work hand-in-hand, recognizing the value of Christian principles in complementing technological development.

Moreover, to truly imprint Christianity's compassionate ethos onto future transport, we must shift our perspectives. Instead of viewing technology as merely a tool for efficiency, we should see it as an entity capable of embodying our deepest-held values of love, care, humility, and collective well-being.

In conclusion, Christianity's imprint on future transportation does not merely align with a dystopian or utopian vision. It proposes a future that balances the marvels of technology with the warmth of love, empathy, and concern for one another. In this holy intersection, we find that the future of transportation becomes not just a question of where we are going, but also how we choose to journey. It implores us to build a world that is sustainable, inclusive, and loving - thereby ensuring that our journey into tomorrow honors our faith and commitment towards one another in today's world. In such a scenario, progress feels less like a cold advancement and more like a warm arrival, where every milestone advances not only efficiency but humanity itself.

Chapter 10. The Mindful Commuter: Personal Reflections on Spirituality and Clean Transit

As we embark on this spiritual and technological journey, the first step lies within nowhere else but our very own selves. Our internal environment significantly influences our actions, and unsurprisingly, our choices surrounding transport.

10.1. Embracing Clean Transit and Spirituality

Our daily commute is more than just a necessity, it's a vessel for reflection and introspection. When we choose clean transportation, we opt for a method that leans towards the fundamental virtues Jesus advocated: selflessness, respect for all creation, and love thy neighbor as thyself.

Many people relate Jesus' message about loving one's neighbor to empathizing with their suffering and reaching out to help in emotional and spiritual matters. However, loving your neighbor can also extend into our influence on the environment. Each of us, as commuters, carry a responsibility towards building a greener planet, a healthier community, and thus, a better future for our neighbors, near and far.

Clean transit introduces us to a mode of transport that's efficient yet low on environmental impact. When we step into a battery-operated bus, ride an e-bike, or simply choose to walk, we're consciously deciding to reduce our carbon footprint. These choices align remarkably well with the ethics and spiritual teachings of Jesus, who emphasised care for our fellow creatures and our shared planet.

10.2. Personal Reflections: A Train Journey

To further illustrate this, let's step aboard a public train powered by a clean, renewable source of energy. As the train departs, it creates an opportunity to reflect and absorb how our actions can resonate with divine wisdom.

As it glides through each station, the train becomes the allegorical backbone of our journey. It isn't just a means of travel; it's a crucible for behavioral change - a transition from the egocentric to altruistic. This public transport mode that runs on clean energy is, in itself, an embodiment of selfless love for our planet and fellow living beings.

The mix of people aboard the train, from different walks of life, presenting multifarious experiences and stories, illustrates the principle of inclusivity. In the divine context, the train's diversity reflects the universal love that Jesus had for all people, regardless of their cast, creed, or social standing. This awareness further fuels the mindful commuting experience, inscribing deep within us an appreciation for our shared equality amidst diversity.

10.3. The Quietude of Bicycling

Another example lies in the simplicity of bicycling. Opting for pedal-power over fossil fuel-driven transport is another act of conscious care for the environment - an act of love.

Cycling offers an opportunity to connect deeply with our surroundings. Each pedal stroke is a manifestation of selflessness, acknowledging our responsibility toward cleanliness and contribution to the overall health of the planet. It's also a chance to process the world around us at a slower pace, allowing for a more mindful and profound appreciation of God's creation.

10.4. Collective Responsibility Driven by Love

As we lean into this new way of commuting - one that is deeply anchored in mindfulness, clean-fuel reliance, and love for our planet and its people - we organically build communities that live by these ethos.

Choosing clean transit is not a path walked alone. It calls for collective action, pushing us to reimagine a world moved by sustainable, inclusive, and loving ways. This reimagining, based not solely on tech advancements but also on Jesus' teachings, is the key to embracing our roles as mindful commuters.

Our transition to clean transit is not just a revolution in technology; it inherently is a spiritual journey. Each of us has the power to translate the divine principles of love, humility, and respect into actions. And imbibing them in our modes of transit, we're not only promoting a sustainable future but redefining what our daily commute signifies.

Each of us carries the potential to be - as in the context of our everyday locomotion - a mindful commuter. Through our choices rooted in love, respect, and humility, we can truly embody the fusion of spirituality and technology, uniting two seemingly distant worlds into one wholesome reality. Embarking on this path, we extend Jesus' teachings from our hearts into our travel maps, creating a divine pattern of transit that supports the planet and each other.

Thus, embracing a commuting lifestyle linked with spirituality instills in us a kind of love that every bit of God's creation deserves. This transformation is more than just a shift towards clean transit; it's a profound movement towards inner peace, collective well-being, and long-lasting holistic changes in our lives.

It is what our world profoundly needs - an integration of spirituality in an essential aspect of our lives, transportation. Through mindful commuting, we preserve not just the world we live in, but also the divine virtues that mold our spiritual selves.

Chapter 11. Riding towards the Horizon: Next Steps in Integrating Faith with Transportation Technology

Our story of convergence begins with understanding and appreciating the multitude of changes in transportation technology currently unfolding before us. This includes, but is not limited to, the proliferation of electric vehicles, advancements in autonomous transportation, growing acceptance of shared mobility and public transportation, and the gradual evolution towards more sustainable and efficient means of transit.

11.1. The Momentum around Electric Vehicles

Electric vehicles (EVs) are transforming the traditional automotive industry at a pace that was unimaginable just a decade ago. With major automakers pledging to phase out combustible engines, the shift towards sustainability echoes Jesus' teachings of caring for our common home, the Earth. As Pope Francis expressed in his encyclical letter 'Laudato Si,' we have a collective responsibility to protect and nourish our planet.

Considering the reduction in carbon emissions and dependence on scarce non-renewable resources, EVs exude this ethos of compassion for the planet and future generations. However, we must act responsibly to ensure raw materials for battery technology are sustainably sourced and that end-of-life procedures are environmentally friendly. It is evident that the rise of EVs presents both opportunities and challenges where our compassion-led approach can drive beneficial practices.

11.2. The Autonomous Revolution

The autonomous vehicle (AV) revolution promises a future of safer, more efficient transportation. The technology removes human error, potentially reducing accidents and saving lives. Yet, it also introduces novel ethical dilemmas.

Applying the principles of Jesus' teachings can help us navigate these dilemmas. A perspective rooted in divine love would prioritize human life above all. The programming of AVs should thus reflect an ethic of care that minimizes harm. Moreover, a concern for social justice would urge us to consider the potential job displacement caused by AVs and find ways to support the affected communities.

11.3. Shared Mobility and Public Transportation

Riding together, whether through public transit or shared mobility solutions, could serve as a modern reflection of communal living and shared responsibility. It reduces individual carbon footprints and promotes a sense of togetherness, helping us understand and empathize with our fellow travelers. This ethos ties back to Jesus' teachings, underlining the potential for a paradigm shift towards more communal transit.

However, efforts should be put into making these solutions accessible to all, regardless of their socio-economic status or geographical location, emphasizing the Christian principle of inclusivity.

11.4. Compassion-powered Innovation: The Next Frontier

Faith can guide us as we shape the transportation of the future. It not only teaches us to consider empathy and compassion in our decision-making but can also inspire us to innovate in ways that promote collective well-being. Whether it is through designing vehicles in a way that makes everyone feel welcome or employing technology to guarantee safety, faith can act as a beacon amidst the uncertainty of change.

As we move forward, faith-based institutions must form alliances with transportation organizations to bring about this compassionate revolution. Initial steps could include facilitating dialogues around the integration of faith and technology, developing educational programs on ethical decision-making, and advocating for policies that promote sustainable transport.

Furthermore, transportation enterprises should increasingly reflect divine love in their corporate identities, integrating compassion, integrity, and sustainability into their mission statements and day-to-day operations. It is vital to remember that technological progression and ethical values are not mutually exclusive, and the synergy of both may provide the solution to many modern-day transportation conundrums.

11.5. A Call to Action

This future is possible, but it requires us all to play our part in driving this fusion of faith and technology. We must recall that faith is not static. It responds to, learns from, and interacts with the world around it. It guides its adherents through an ambiguous future filled with both exciting potential and daunting challenges.

Similarly, we must approach technology positively, recognizing its capacity for good while remaining mindful of its potential pitfalls. We must remember that technology, like faith, is a tool. It is a means to an end, and it can be wielded in ways that reflect the divine love at our core. If we can do this, if we can truly integrate faith and technology into our transportation reality, then we are indeed riding towards the horizon and into a future where our travels are imprinted with the indelible mark of divine love.